Primary Science Equipment

Contents

The purpose of this book	4
Developing a whole school policy on science equipment	5
Developing a curriculum plan for equipment in science	6
The role of equipment in primary science	7
The importance of measurement in science	8
Progression in choosing and using science equipment	9
Science resources should be ... the responsibility of everyone	12
Responsibility	13
Helping hands	14
Storing equipment	15
Equipment list	18
Equipment audit list	43
Small equipment and consumables	47
Photocopiable equipment labels	49
Useful addresses	67

The purpose of this book

This book aims to provide:

- a list of basic science equipment which meets the needs of science in the primary school;
- useful information about individual items of equipment;
- safety information;
- a rationale for the development of the use of equipment throughout the school;
- overhead projector masters for use in policy development meetings and INSET;
- suggestions for methods of storing science equipment;
- photocopiable checklists for auditing and purchasing science equipment;
- photocopiable labels for equipment.

The material in this book can be used:

- to assist schools in the process of reviewing science equipment;
- to enable schools to note gaps in basic provision;
- as a reference for staff responsible for science in school;
- to help in the development of a school policy towards equipment provision and use;
- to assist in developing suitable methods and procedures for organising and using equipment across the school;
- to create labels for equipment.

Developing a whole school policy on science equipment

Resourcing primary science takes time, money and effort. It is therefore important to give considerable thought to the management and development of science equipment so that the school gets value for money and children and teachers are adequately supported.

When developing or updating a school policy on primary science equipment the following should be considered:

- How will the staff be encouraged to take ownership of this policy?

- What are the equipment needs of the school and individual members of staff?

- How will staff be encouraged to adopt new approaches to organisation and use of equipment?

- What are the long-term plans for developing the policy? Will dates and targets be set?

- Which storage system does the staff prefer?

- What strategies will be developed to encourage children to be responsible for equipment?

- What are the short-term purchasing priorities? Can 'friends of the school' provide some materials?

- What are the long-term purchasing priorities? Is external funding required?

- What are the staff development needs? For example, INSET on computer sensing equipment, development of whole school policy, classroom strategies for developing pupil responsibility for equipment.

- Who will be involved in organising and looking after equipment in school?

- Will the policy encourage parents and pupils to organise and care for equipment?

Developing a curriculum plan for equipment in science

Once a school policy on resourcing primary science has been agreed, a statement should be written into the school science policy. Further information should then be provided in curriculum documentation, such as the science coordinator's file.

This could include the following:

- A copy of the school policy statement relating to resourcing primary science.

- A rationale for the organisation and development of science resources.

- An audit of science equipment.

- An indication of the location of science resources.

- A log of previous expenditure on equipment.

- Current resource priorities and spending in relation to the audit of equipment.

- Projected resource priorities and finance plan for the next 2–5 years.

- INSET provision relating to equipment and other resources in science.

The role of equipment in primary science

Equipment is an important feature of primary science and has a role to play in developing how children think and work scientifically. Although some items of equipment can be produced by children or made by teachers or friends of the school, primary science has some specific requirements for robust and precise equipment. As children get older they should begin to appreciate the importance of equipment in science.

For example children should:

- Appreciate that equipment must be used with care and safely.

- Become aware that there are specific tools to help them in science, ranging from hand lenses to hanging masses.

- Understand that there are different kinds of science equipment that have specific jobs, e.g. Newton meter.

- Develop the ability to decide when equipment is necessary, to choose appropriate equipment and use it carefully and accurately.

- Understand that measurement is important in science and that equipment can help to provide accurate measurements.

- Appreciate that equipment used in the classroom often has an equivalent in real life, e.g. a market gardener might use a pH tester for soil and a humidity sensor in greenhouses.

- Begin to appreciate that the skills they develop using equipment in science can also be used in everyday life.

- Be aware that there is often a range of one type of equipment, for example thermometers, and that they have to decide which type of thermometer is best for the job.

The importance of measurement in science

Measuring equipment has a central part to play in many primary science activities, particularly field work and science investigations. As children develop their scientific ability they should rely increasingly on the use of measuring equipment.

Measurement can:

- *Encourage children to collect more accurate information in a range of contexts, from insulation to stream pollution.*

- *Lead to the more frequent use of numerical data, which is important for illustrating patterns in results.*

- *Generate numerical data and thus provide better opportunities for pupils to make comparisons.*

- *Provide evidence of results which can be used to support and justify conclusions.*

- *Allow pupils to ask questions and challenge information offered by other children.*

- *Provide opportunities for children to develop an understanding of reliability and validity.*

- *Help children to understand how data can be viewed differently by different people.*

- *Help children to develop the confidence to challenge data.*

Progression in choosing and using science equipment

The development of children's understanding of the use of equipment should be planned across the primary years. Expectations for year groups should be made explicit in terms of which equipment is introduced and the level of understanding of the different types of equipment, how it can be used and safety and care issues. Gradually the responsibility for choosing and using equipment should move from the teacher to the child. The following provides a suggested framework for progression in choosing and using science equipment across the primary years.

Level	Choosing equipment	Using equipment	Safety and care
Nursery	• The teacher provides a range of equipment for children to use in both real-life and play situations. For example, kitchen scales when cooking, 'elephant' tape measures for role play. • The teacher should use and explain equipment to children and allow children to try out equipment for themselves.	• Children should be introduced to using equipment such as kitchen scales when cooking. • A range of equipment, from hand lenses and scales to tape measures, should be available for children to choose from in their 'play' situations.	• The teacher needs to introduce children to using equipment safely. • Children need to be taught routines for getting equipment out and putting it back in the correct place. • Pictures and/or silhouettes can be used to help children locate equipment. • Children need to develop an appreciation of how to share items of equipment.
Reception	• The teacher is responsible for choosing equipment. • The teacher demonstrates to children how to use equipment. • Children are encouraged to suggest what kind of equipment they could use. • Children begin to choose from a limited range of equipment provided by the teacher.	• Children follow teacher instructions. • A teacher or adult oversees the use of equipment. • Some support for children with manipulative difficulties is often required.	• The teacher needs to introduce children to using equipment safely. • Children need to be taught routines for getting equipment out and putting it back in the correct place. • Children require the support of pictorial labels to help them return items to the correct place.

Progression in choosing and using science equipment

Level	Choosing equipment	Using equipment	Safety and care
Year 1	• Children become more familiar with a wider range of equipment. • With teacher support children choose equipment. • Children are encouraged to remember which equipment they have used before, when making decisions.	• Children might need to be reminded of how to use equipment. • Teacher keeps a 'weather eye' on children using equipment. • Teacher begins to share responsibility for using equipment with children, gradually allowing children more independence with equipment.	• The teacher should ask children how equipment should be used. • Children should be asked about using a piece of equipment safely, and told if they cannot remember. • The teacher should begin to give over responsibility to children for getting some items of equipment out and putting equipment away safely and tidily. • Children should also begin to tell the teacher when equipment is damaged or pieces are missing.
Year 2	• Children should be familiar with the basic range of equipment and be able to choose appropriately with only limited help from the teacher.	• Children should be able to use familiar pieces of equipment without assistance. • Children should begin to appreciate the need to use measuring equipment accurately.	• The teacher should expect that children know how to use common pieces of equipment safely. • Children should be responsible for collecting and putting equipment away. • Children should also be able to tell the teacher when equipment is damaged or pieces are missing.
Year 3	• Children become increasingly independent when choosing equipment. • The teacher's role moves towards checking and validating children's choice of equipment. • New equipment still needs to be introduced by the teacher, with explanations and demonstrations of how to use it.	• Children should be able to use basic equipment with confidence. • Children should know that accuracy is important when using some pieces of equipment. For example, measuring equipment. • Children should be able to use measuring equipment with increased accuracy.	• Routines should be established for collecting and putting away equipment. • The teacher should expect only to supervise the safety and care of equipment. • Children should be able to suggest safety precautions before using certain pieces of equipment.

Progression in choosing and using science equipment

Level	Choosing equipment	Using equipment	Safety and care
Year 4	• Children should require little direction from the teacher in relation to choosing familiar equipment. • Children should be able to choose from across, as well as within, ranges of equipment. • Children should be introduced to the use of computer sensors in science and taught how to use them.	• Children should be able to use some equipment without adult intervention, except where equipment fails. • Children should begin to problem-solve if equipment fails to work and suggest reasons to the teacher. • Children should know that accuracy is required when using measuring equipment.	• Children should be able to suggest rules and routines for accessing and using equipment. • Children should be able to take some responsibility for monitoring equipment. • Children should be familiar with basic safety rules relating to the use of common equipment.
Year 5	• Children are able to choose appropriate equipment from within a range. For example, capacity containers, timers. • Children should require minimum adult support except where new equipment, for example a Newton meter or pH probe, is introduced.	• Children should be able to use equipment in a confident manner. • Children should demand the accurate use of equipment themselves. • Children should begin to appreciate when repeated measurements are required.	• Equipment should be the responsibility of the children using it. • Children should indicate when equipment is broken or consumables have run out. • Children should be required to indicate to the teacher how to manage safety precautions when using certain equipment, for example night lights.
Year 6	• The teacher should expect children to be able to choose the most appropriate equipment for the task. • Children should be able to suggest substitute equipment if the most appropriate is not available. • Children should be confident in choosing and using computer sensors where appropriate.	• Children should be able to use equipment in a confident manner. • Children should know why accuracy is required when using equipment. • Children should demand that equipment is used in an appropriate and accurate way. • Children should be able to set up computer sensors for use in activities.	• Children should recognise the safety implications of using certain equipment. • Children should be able to offer their own strategies for managing safety in activities. • Children should appreciate that they are responsible for their own safety and that of other children.

Science resources should be:

IN SUITABLE QUANTITIES

Some equipment requires multiple purchases, for instance of batteries, bulbs, wire and crocodile clips. Other larger and more expensive items of equipment do not. It is important to consider how science is managed throughout the school; this will determine the purchasing policy.

UP TO DATE

New equipment is constantly coming on to the market, for example computer-sensing equipment. It is important to occasionally browse through catalogues to keep up to date with information, so that finances can be planned to allow the purchase of new equipment. Keep an on-going notebook to aid end-of-year ordering.

COUNTED OUT AND COUNTED IN

If equipment is stored in containers then the number of items inside should be indicated on the label outside. When equipment is used the borrower will know how many items there should be inside. A quick count at the end of an activity helps to recover equipment and reduce losses.

CHILD-FRIENDLY

Some equipment is designed to be colourful and robust, interesting to look at as well as to use. Before purchasing equipment ask yourself whether it looks inviting for children. Would you want to use it?

SAFE

Equipment should be checked on a regular basis to ensure that it is in good working order.
The ASE book *Be safe!* provides important information about safety in primary science.

WIDE RANGING

Primary science requires that children should develop the ability to choose appropriately from a wide range of equipment. They should be able to recognise 'fitness for purpose', that is which is the best piece of equipment for the job. This requires a good range of equipment.

APPROPRIATELY STORED

Storage should reflect the needs of both children and staff. Equipment should be stored in an attractive and imaginative manner. It could be stored according to topics or in alphabetical order using the labels provided at the end of this book.

WELL MAINTAINED

Nothing is more frustrating than finding equipment that does not work because it is broken or, for example, batteries are flat. Encourage users to report breakages, etc., verbally or to use a system where a large, brightly coloured label can be stuck on a box to indicate the need for maintenance.

LABELLED

All storage facilities should be carefully labelled with a picture of the equipment as well as the name to support children who have difficulty reading. Use the labels in this book to help children locate equipment and also so that they know where to return it.

ACCESSIBLE

Resources should be stored at a level that enables children to reach them easily and safely. Children should be able to see inside containers and/or see labels indicating with both words and pictures (or silhouettes) the equipment inside.

The responsibility of everyone

Responsibility

Children should have responsibility for identifying which type of equipment they need for their activity, and getting it out. Then children should be responsible for putting it away.

Responsibility should be shared. It is the only way that others, both staff and children, develop a sense of ownership towards equipment.

The teacher should teach children how to use equipment safely. Then children should be given increasing responsibility for using equipment in a safe and proper way, *although overall responsibility always remains with the teacher.*

Being willing to share equipment and wait one's turn are important and responsible attitudes to develop in children.

Routine is important for developing a responsible attitude. Demanding that children collect and return resources each lesson is important and providing opportunities for parents or helpers to check resources on a regular basis.

The expectation that resources should be taken out and returned in an orderly manner should apply to everyone, children and adults alike.

Be explicit about who is responsible for what. Offer guidelines, for example 'If the batteries do not work – leave a sad face sticker.'

Offer strategies to encourage children to develop responsible attitudes towards science equipment. Develop common whole school approaches which staff incorporate into lessons.

Helping hands

- Children and adults can be trained as equipment monitors. Delegate responsibility for the everyday running of resources.

- Staff could be given the task of creating equipment lists for each of the topics they teach.

- Do make sure that the school Governors and parents know if you are developing school science resources. They might be able to offer help.

- Lists of consumable materials, such as corks, plastic containers, etc., can be given to classes in the school to take home. Each class then becomes responsible for collecting the items on their list.

- Give the job of checking the equipment on a regular basis to someone else, for example older children, parents, 'friends of the school'. Ask them to make a note of broken equipment and when items have run out.

- Quality containers are important but can be expensive. A local business or parents might donate suitable containers for storing equipment.

- Keep a 'running low' list for people to fill in: then you won't have to keep going through the equipment to find out which items need replenishing.

- If you delegate be specific; make sure each person knows exactly what to do and how long it should take. Monitor progress and outcomes.

- Some industries are pleased to link with schools. Although they might not provide money, some run 'recycling' schemes where materials they do not want, are offered to schools.

- Parents and friends of the school can be asked to set up and organise a resource area. Make sure that you have regular meetings to review progress.

Storing equipment

Central storage

In this method all science equipment is stored centrally. For example:

- all equipment for the whole school in one area, such as a designated room, corridor or store room;
- separate central stores for infant and junior phases.

Advantages

- Easy access to equipment.
- Everyone knows what is available.
- Where equipment is stored on shelves which are labelled it is easy to see which items have been borrowed.
- Checking and maintaining equipment can be carried out easily and in an efficient way.
- Everything is at hand; there is no need to search throughout the school for items of equipment.

Limitations

- Requires part of the school to be set aside as a resource area.
- Requires cupboards and/or shelving and suitable containers.
- Careful labelling is crucial so adults and children can locate equipment quickly.
- A signing out system is necessary to keep track of equipment.
- Science planning needs to be monitored throughout the school to ensure that there is sufficient equipment, particularly where whole school topics are used.
- Duplication of some equipment may be necessary to manage demand.

Topic boxes

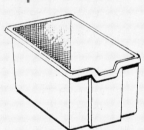

Equipment is stored in science topic boxes such as forces or electricity. Topic boxes are borrowed by classes for a set period of time.

A school might organise the topic boxes in one of the following ways:

- one electricity box for the whole school;
- three electricity topic boxes – one for reception/key stage 1, one for lower key stage 2 and one for upper key stage 2 – each containing equipment appropriate to the key stage.

Advantages

- The equipment necessary for a topic is in one box.
- The box can be taken to the classroom, keeping everything together.
- An equipment checklist can be attached to the box so that the contents can be checked easily.
- Teachers can add equipment and teaching materials to each box.
- Pupil and teacher curriculum materials to support the topic can be kept in the box.
- The class using the topic box can be given responsibility for checking and maintaining the box and reporting breakages and losses.

Limitations

- Storage boxes must be strong and carefully packed to avoid damaging equipment.
- Boxes need to be checked at the end of each topic and where necessary equipment replaced or repaired.
- Items of equipment in common use across topics have to be duplicated.
- This system requires forward planning of science topics across the school and for at least two years in advance.
- The contents of the box could dictate the science taught and might limit alternative and perhaps more creative approaches to science.
- Measuring equipment has to be duplicated.

Storing equipment

Individual class resource

Each class is provided with a set of equipment which should cover most of the science to be taught that year.

Advantages	Limitations
• Children have immediate access to science equipment. • Checking and maintaining equipment is delegated to each class. • Neither the children nor individual class teachers have to spend time collecting and organising science equipment. • Once set up the system can be in place for several years. • Teachers can provide a list of equipment which needs to be replaced either each term or once a year.	• Duplication of equipment is necessary. • School topics have to be planned at least one or two years in advance. • Teachers have to organise and monitor the storage of equipment in their own classrooms. • The type of equipment provided might limit science activities.

Basic equipment kit for each class

Each class is given a set of basic science equipment, such as hand lenses and measuring equipment. This equipment kit is to accommodate spontaneous science activities in the classroom and basic measurement in activities. Science equipment related to the main teaching topic, for example electricity, is borrowed from a central store or a topic box.

Advantages	Limitations
• Children always have immediate access to items such as hand lenses. • Children and teachers can engage in science which arises from spontaneous interest or an event. No one has to wait for another class to return equipment. • Equipment is easy to monitor if labelled and numbered.	• It can be expensive to provide each class with a basic set of equipment. • This system requires a separate store of equipment elsewhere in the school. • Some classes might be more careful with their equipment kit than others. • Difficulty of how to sanction a class which repeatedly breaks or loses items of equipment.

Storing equipment

Measuring equipment trolley

A measuring equipment trolley is simply a trolley with shelves and trays in which measuring equipment is stored for use in science activities. The items are numbered and labelled and the trolley either stored in the classroom (one trolley per class) or moved from class to class as the need arises or the class is timetabled for science.

Advantages

- The equipment is portable and can be moved from class to class.
- Classes can be allocated the trolley for a certain period, such as a half day or a whole day.
- It is an efficient use of equipment.
- It provides good value for money, ensuring that a wide range of equipment is available for all children.
- Separate equipment trolleys can be provided for key stages 1 and 2.
- The measuring equipment can be used to support other curriculum areas.

Limitations

- This can be costly to set up.
- It requires multiple sets of equipment where more than one trolley is necessary.
- A portable trolley is problematic in a multi-storey school.
- Careful monitoring is needed.

Equipment list

Equipment	Types and uses	Safety and care	Approximate cost/£

The environment

Equipment	Types and uses	Safety and care	Approximate cost/£
Anemometer	• Used to measure wind speed. • Hand-held anemometers are used for localised measurement. • Field anemometers are for continuous use in a fixed position; they are weather-proof. • Anemometers are useful when studying the conditions of a particular habitat, e.g. field, hedgerow.	• Store with care to avoid damaging the cups. • Use and site according to manufacturer's instructions.	10–20
Aquarium	• Used for observing and housing living organisms, i.e. fish, stick insects, snails, spiders. • A useful transparent container for water-related investigations such as floating and sinking, passing light rays through water.	• If used for animals ensure that the teacher and children have sufficient knowledge about maintenance of species involved. • Plastic aquaria should not be carried by the lip; it can break off. Teach children to carry the aquarium with their hands underneath. • Ensure that electrical appliances are not placed near the aquarium. • The plastic does scratch so care should be taken not to place large stones or objects in the aquarium.	6–10
Barometer	• Measures air pressure, used to predict weather patterns.	• Use according to manufacturer's instructions. • Can be purchased individually or as part of a weather station. • Avoid buying mercury barometers; ask for the aneroid type.	30–45
Binoculars	• Allow closer observation of wildlife in natural habitats.	• Usually an expensive item; **must** be handled with care. • Ensure neck strap is worn to avoid children dropping the binoculars. • When not in use store in case. • The optics should be cleaned with a suitable cloth. • Insist that children **never** look directly at the Sun with or without the binoculars.	20–30

The environment

Equipment	Types and uses	Safety and care	Approximate cost/£
Compass, magnetic	• Used to locate magnetic pole and related points of the compass. • Simple plotting compasses are useful for magnetism work and basic direction work. • Can be used for work linked to shadows, and apparent movement of the Sun. • Other compasses are more complex and used for orienteering work.	• Do not store near strongly magnetic materials. • Except for 'field' type orienteering compasses, ensure that they are always used on a level surface to allow free movement of the needle.	1–2
Flower press	• For the preservation of leaves and flowers.	• Ensure that the minimum of specimens are taken. • Ensure that no rare or uncommon wild species are used. • Avoid poisonous and other plants likely to cause allergies. See the *ASE* book *Be safe!*	5–10
Gardening tools	• Spade – for general digging. • Dutch hoe – for breaking up surface soil and weeding. • Fork – for deeper soil. • Rake – for creating level surface and clearing debris. • Hand trowel – for localised digging and planting. • Hand fork – for localised digging and planting. • Secateurs – for cutting back and pruning plants. • Watering can. • Garden spray – for plant misting. • Wheelbarrow – for gardens with larger area.	• Always store gardening tools in a dry and clean condition. • Purchase child-size tools; adult tools are too large and therefore dangerous. • Ensure that children are shown how to use and handle tools correctly and safely.	3–20
Globe	• For use in teaching 'Earth in Space', e.g. night and day. • Range from small globes on stand to inflatable globes.		15–40
Hygrometer	• Paper – for measurement of relative humidity of local environment. • Wet and dry bulb – the temperature around a dry bulb is compared with the moistened bulb; hence relative humidity is found.	• Store papers in a dry place.	5–10

The environment

Equipment	Types and uses	Safety and care	Approximate cost/£
Light meter/sensor	• Measures level of light. Displays light reading either as numbers or letters. • Can be a hand-held battery operated meter or a computer sensor. • Used in field work, for example taking light readings in different parts of a woodland habitat.	• As with all field equipment ensure that children use with care. • Check equipment out and in. • Some photographic light meters operate by solar power.	15–20 for hand-held meter 40–50 for computer sensor
Litmus paper	• For the measurement of the alkalinity or acidity of a liquid. *Red – alkali* *Blue – acid* *Neutral*	• Store in a dry place. • Use according to instructions.	1–2
Nets	• For collection of live air and water animal specimens. • Small household sieves can be used for shallow pond dipping. • Long-handled nets are required for deeper ponds.	• Only catch a minimum number of specimens. • Place in a suitable white container or tray of water immediately for observation. • Handle animals with care. • Release animals into habitat as soon as possible after observation. • Clean nets and sieves after use.	2–22
pH meter/probes	• For obtaining a pH measurement of solutions and soil. • Provides instant reading of alkaline/acid soil. • Portable, pocket, battery-operated meters can be purchased for fieldwork.	• Some cannot be used in water.	6–15
pH papers	• For measurement of pH in solutions. • Less accurate than pH meters.	• Need to be kept dry.	1–3
Plant propagators	• Can be purchased in a range of sizes for growing seedlings and sensitive plants. • Some propagators are thermostatically controlled.	• Use according to instructions. • Take particular care if using electric propagator. • Children can make their own propagators using cut-down clear plastic containers or plant pots with plastic bags over the pot. (Care required when using plastic bags.)	3–50

The environment

Equipment	Types and uses	Safety and care	Approximate cost/£
Pooters	• For collecting very small animals. • The animals are sucked up into a container via a tube.	• Use with care to avoid damaging animals. • Only very small animals can be collected. • The tube mouthpiece should be disinfected after use. See the ASE book *Be safe!*	2–3
Quadrats	• Transparent sheet marked in metric grids and radials or squares with string grid. • Used for counting items in a specified area. • For work on habitats, particularly fields, hedgerow and woodland. • Often collapsible squares; therefore portable. • Some can also be used on vertical surfaces, e.g. walls, hedgerows.	• For field work give each child one piece of equipment to look after. • Each child is then responsible for that piece of equipment. Before leaving a site ask children to hold up their piece of equipment. It soon becomes obvious if something is missing. • Simple quadrats can be made in Design Technology or by parents.	4–10
Rocks and fossils	• Sets of rocks and fossils can be purchased for use in identification and classification activities. • Rock collections are also useful for work on the properties of materials and to illustrate that many materials used originate from rocks, e.g. coal, talcum powder.	• Store carefully to avoid damage. • If not labelled use nail varnish or enamel paint to number specimens. • Keep a record of number and type of rocks. • Let parents know that you are developing a rock collection. Samples might be donated; do ask for information to be included with any donations. • Small samples can be stored in egg boxes.	15–50
Rain gauge	• Used for the collection of rain water. • Also used to measure amount collected.	• Needs to be sited in an open area away from overhanging roof or trees.	4–10
Soil meters	• Moisture – able to measure moisture content at various depths in different soils. • pH – a specially designed electrode that can be used within the soil to measure pH. • Temperature – thermometer able to measure soil temperature at various depths and designed to be pushed into soil without damaging the thermometer.	• Always clean and store carefully after use. • Although robust, care is still needed when using equipment.	5–15

The environment/Ourselves

Equipment	Types and uses	Safety and care	Approximate cost/£
Soil sieves	• These can be purchased separately or as part of a set of 3–4 nestable meshes. They allow for the separation of soil particles by size. • Kitchen strainers are a cheaper alternative.	• Use and store with care to avoid damaging the mesh. • Do not force particles through. • Use with dry soil. • Clean using water or leave to dry and use a **soft** brush to clear soil.	2–40
Weather board	• A board for recording weather observations.		15

Ourselves

Equipment	Types and uses	Safety and care	Approximate cost/£
Anatomical models Various types described below. **Human torso**	• These help children to appreciate the inner workings of the body. Many anatomical models have removable parts to illustrate the different layers and positions of body parts, particularly internal organs.	• Most anatomical models are too expensive for purchase by primary schools. However, sharing the cost within a cluster group is possible. Alternatively, some secondary schools are willing to lend this type of equipment for short periods of time. • Where there are removable parts provide a list which numbers and names the parts to ensure the correct number is in place after use.	
	• Shows the internal organs of the body.	• Ensure that removable parts are taped on to the model when not in use.	20–40
Ear	• Shows external and internal parts of the ear. Some models have removable parts.	• Ensure that removable parts are taped onto the model when not in use.	40–50
Eye	• Shows the external and internal parts of the eye.	• Ensure that removable parts are taped on to the model when not in use.	40–50

Ourselves

Equipment	Types and uses	Safety and care	Approximate cost/£
Heart	• Shows the external and internal parts of the heart.	• Ensure that removable parts are taped on to the model when not in use.	40–50
Human anatomy	• Shows the internal organs of the body, muscles, skeleton. Usually the whole of the human body.	• Ensure that removable parts are taped on to the model when not in use.	30–70
Inside-out tunics	• Tunics for children to wear showing major organs or skeletons. Some have removable parts and matching Velcro body-part labels.	• Ensure that removable parts are on the tunic when not in use and store in a box or a bag.	40–50
Skeleton	• Skeletons can vary in height and detail. Most have stands, some have boxes and many are made to ensure that joints articulate in the correct way.	• Store carefully, either in a box or on its stand with a cover.	30–50
Tooth	• Illustrates a set of teeth or individual tooth	• Ensure that removable parts are taped on to the model when not in use.	30–130

Ourselves/Sound

Equipment	Types and uses	Safety and care	Approximate cost/£
Dental charts	• These can be obtained from local dentists and photocopied so that children can complete them to show fillings, cavities, extractions. • Ask the dentist for copies of charts already completed, with, of course, the names removed.		Free
Heartbeat counter	• A pulse sensor which provides readings of heartbeat. • These range from wrist-watch pulse-rate monitors, using a sensor which fits over the finger, to more expensive units which use an ear clip and have a range of functions.	• Care should be taken that children do not over-exert themselves when carrying out activities to record changes in heartbeat with change in activity level.	10–100
Stethoscope	• A device for listening to sounds, particularly the heart beating.	• Stethoscopes vary in quality and certainly work less well if used over several layers of clothing or in a very noisy area!	5–10
X-rays	• Can be obtained from local clinics or hospitals. The patients' names will have been removed.	• Store in large envelope to prevent damage.	Free

Sound

Equipment	Types and uses	Safety and care	Approximate cost/£
Musical instruments	• Use instruments from the school's music collection. • A collection of instruments can be purchased from catalogues, which usually provide a limited range of instruments, e.g. tambourine, cymbals, maracas, kazoos.	• All instruments should be handled with care and stored in an appropriate manner. • All instruments should be returned to the music store. • Teach children that they should not make sudden or long bursts of very loud sounds; these can damage the ear. • Musical instruments can be made in Design and Technology.	40–60 per set

Sound

Equipment	Types and uses	Safety and care	Approximate cost/£
Plastic tubing	• For use in making voice tubes (telephones) and stethoscopes. • Different diameter tubing can be purchased by the metre.	• Where children talk down the tube, make sure that they do not make excessively loud noises which could damage the listener's ear.	0.20–2.00
Plastic funnels	• For use in making voice tubes and stethoscopes.	• Where children talk into the funnels, make sure that they do not make excessively loud noises which could damage the listener's ear.	0.30–2.00
Pipes (metal)	• Range of metal pipes which can be used to illustrate changes in sound according to material and length/width of pipe.	• If pipes are cut down ensure that all rough and sharp edges have been smoothed.	
Slinky	• To show compression waves when teaching sound.	• Do not stretch slinky to its full extent.	1–5
Sound-level meter	• A portable meter which measures loudness of a sound. • Computer sensors which measure sound are also available.	• Children should not make loud or sudden sounds close to the ear.	40–50
Tuning forks	• Used to teach vibration and pitch. • These can be purchased singly or as a set. • Each fork is set at a certain pitch. If buying a limited number ensure that there is a range of pitch. • Collect a range of forks with different 'notes' so that tunes can be played.	• Do not hit forks on a hard surface; this can damage the fork. • Store carefully.	Single 2–5 Set 30–40

Electricity

Equipment	Types and uses	Safety and care	Approximate cost/£
Batteries (dry cell)	• Standard dry cells are normally used for primary science. • Batteries range in size from 1.5 V to 4.5 V. • 4.5 V flat-shaped battery with knife terminals will allow children to attach crocodile clips. • Round and cuboid batteries require battery holders and battery clips.	• Never mix different types of batteries. • Ensure that voltage of battery matches the requirement of the component, e.g. 3.5 V bulb with 4.5 V battery. A 1.5V bulb with a 4.5 V battery will burn out. • Rechargeable batteries are not normally suitable for use in primary schools. • Rechargeable batteries can discharge rapidly, overheat and often melt the surrounding battery holder. • For further information see the ASE book *Be safe!* • Alkaline/rechargeable batteries should not be used for ordinary circuit work.	0.80–1.80
Battery holders	• A wide range is available so ensure that those chosen will take the number and size of battery used. • For connection into circuits some require battery snaps.	• Do not use rechargeable batteries as they can melt the holder. • Do not leave batteries in holders in case they leak.	0.30–0.80
Bells	• Bakelite casing, larger than buzzers.	• If the bell is correctly connected but does not work the hammer arm might require adjustment. This is a simple procedure but does require a small screwdriver.	5–10
Bulbs (lamps)	• Normal threaded bulbs are known as MES cap. • Bulbs (lamps) range from 2.5 V, 3.5 V to 6 V.	• Teach children to handle bulbs with care to avoid breakages. • Match the voltage of the bulb to the battery. • Place broken bulbs in a tin labelled 'Broken Glass'. Do not place in the waste paper bin.	1.75 per 25
Bulb holders (lamp holders)	• Normal lamp holders with screw fitting are known as MES batten type holders. • MES clip-on type clip on to items such as lolly sticks; these are useful in model making.	• Check connections when a bulb does not work; usually the bulb is not screwed in properly.	1.50–2.00 per pack of 10

Electricity

Equipment	Types and uses	Safety and care	Approximate cost/£
Buzzers	• When connected in a circuit make a buzzing sound. • Small and easily hidden in models.	• These need to be connected the right way round. If the buzzer does not work reverse the connection.	6–7 per pack of 10
Crocodile clips	• Standard clips require connecting wire to be attached. • Leads with insulated crocodile clips can be purchased.	• Some clips are too stiff for young children or those with manipulative difficulties. Check before purchasing or purchase a range of clips. • Crocodile clips with leads can become tangled; use large pieces of card and attach clip to one side of the card and the other clip to the opposite end of the card. • Alternatively, clip ends of leads together so that they do not become tangled up with other leads.	0.90 per pack of 10 (clip only) 1.50–2.00 per pack of 10 (clip and lead)
LEDs (light emitting diodes)	• LEDs are used for the lights on a computer, cooker, TV, etc. • Small lights in red, yellow or green with prongs to attach clips. • Some LEDs flash on and off.	• LEDs connect only one way; if bulb does not light try reversing the connection. • They are quite fragile; teach children to use with care. • They are easily lost; keep LEDs in a separate container from other bulbs.	0.10 each
Motors	• Motors are available in different sizes. • By reversing the connection the movement of the motor can be reversed. • Motors can be used with model propellers.	• Insist that children do not leave motors running; this drains the battery quickly. • A motor pulley might be required to fix items on to a motor.	0.25–0.50 each
Screwdrivers	• Smaller than conventional screwdrivers for use on bells and crocodile clips.	• Teach children how to use a screwdriver. Always wind wire around the screw in the direction the screw is turned. This means that the wire is screwed in as the screw is. • Children should take care when using screwdrivers.	0.50–1.50 each
Switches	• Bell push switch. (1) • Press/push button switch. (2) • Reed switches – the switch opens and closes as a magnet is passed by the switch. (3) • Slide on/off switch. (4) • Toggle switch. (5) • Throw switch – also known as knife switch. (6) • Tilt switch – opens and closes circuit as it is tilted. (7)	• Tilt and reed switches are small components and easily lost. They should be stored carefully in suitable containers. • Reed switches are fragile and also have small magnets. Children should make sure they do not lose magnets. • Some tilt switches contain mercury, but non-mercury ones can be purchased.	0.18–1.20

Electricity/Magnets

Equipment	Types and uses	Safety and care	Approximate cost/£
Wire	• Multi-strand wire is usually used since it is less likely to snap and cause circuit failure. • Most wire used in primary science is insulated copper multi-strand 3 amp wire. • It can be purchased on 100 m reels. • Resistance wire is also useful for work on resistance in circuits. • Clips and wires can be purchased which are already made up.	• Use good wire strippers to cut and strip wire. • Wire on reels allows children to cut only what they require. • Where crocodile clips are purchased already attached to wire the length is standard and sometimes too long, particularly for model making.	6.50–7.50 per 100 m reel
Wire strippers	• Metal wire strippers, although cheaper than automatic wire strippers, are more difficult to use. • Purchase automatic wire strippers which are adjustable according to the wire. Children find these easy to use.	• Metal wire strippers can be hard to use and hurt children's hands.	1.30–1.70 metal 2.75–3.50 automatic

Magnets

Equipment	Types and uses	Safety and care	Approximate cost/£
Alnico magnets	• These are bar or horseshoe magnets which are very strong and excellent for teaching how magnets can attract and repel each other.	• As with all magnets they should not be dropped or banged about as this can reduce the magnetic force of a magnet. • Although very strong these magnets are brittle and if dropped can break, forming two smaller magnets. • All magnets should be stored with keepers (small metal bars placed at the end of magnets). • All magnets should be stored carefully. They should not be placed directly next to each other as this can affect their magnetic field.	6–8 a pair
Bar/horseshoe magnets	• Come in a range of sizes usually made from chrome steel. • These are adequate for most activities but they cannot compare with the more expensive and stronger Alnico magnets.	• Being weaker magnets, these lose their magnetic properties more quickly than Alnico magnets. • Do not bang or drop these magnets. Store with keepers. • When bar magnets have lost their strength they can be re-magnetised using a solenoid (cylindrical coil used to re-magnetise magnets) which can be borrowed from a local secondary school.	2–4 a pair
Ceramic magnets	• Come in a range of shapes and sizes often sold in packs.	• As with all magnets they should not be dropped or banged about as this can reduce the magnetic force of a magnet.	0.30–1.50 each 10–20 for pack of 300 small pieces

Magnets

Equipment	Types and uses	Safety and care	Approximate cost/£
Floating (ring or polo) magnets	• Round magnets with a hole in the centre. • Excellent for illustrating attracting and repelling. If 'threaded' on doweling rod, one on top of the other, these magnets can be made to 'float' on top of each other. • The poles of these magnets are on the faces of the ring.	• If using doweling ensure that the point of the doweling has a piece of Blu-tack or some other safety device attached to prevent any possible danger to a child's eye.	2–4 a set
Horseshoe magnets (infant)	• Large magnets which are easy for young children to handle; often in the shape of an animal. • Stronger horseshoe magnets can lift several kilograms.	• Magnets should be stored with keepers to help retain magnetism.	1–5 each
Iron filings	• Can be purchased as either fine or coarse. • Can be used to illustrate magnetic fields. • Can be used to create iron filing pictures.	• Avoid fine iron filings since they can be inhaled. • Iron filings should be used in a sealed container, such as a sealed Petri dish or a box with a transparent lid.	0.50–1.00
Iron filings bubble	• Iron filings encased in plastic bubble. • Can be used to create pictures and patterns.	• Safe way to handle iron filings. • Alternative is to place filings in a sealed container, such as a sealed Petri dish or a box with a transparent lid.	0.50–1.00
Magnetic marbles	• Brightly coloured plastic marbles with small bar magnet inside. • Useful for challenging children to explain what they think is happening when marbles attract or repel each other.	• Count them out and in; these magnets are easily lost.	1.30–2.00 per pack of 20
Magnetic rubber strips and sheets	• Rubber strips which are magnetic. • Can be cut to size.		1–4 for packs of 4 rubber strips 5–6 for 10 sheets

Magnets/Light

Equipment	Types and uses	Safety and care	Approximate cost/£
Magnetic tape	• Looks like a roll of tape and can be purchased in different widths. • Can be cut to size.	• As with all magnetic material make sure that children keep magnets away from their watches.	6–10 per roll
Magnetic wands	• Plastic casing with magnet inside one end of the wand. • Good for magnetic work with young children, brightly coloured and easy to hold.	• Explain to children that they should not bang the wands since it de-magnetises them.	3–10 per set
Metal discs	• Discs made from magnetic and non-magnetic metal. • Each disc has the name of the metal printed on the reverse.	• Store in a container which lists the number and name of each of the discs.	2–10 per set
Soft iron rods	• Used to create electromagnets by winding wire around the rod and connecting in an electric circuit.	• Electromagnets drain the battery if left on.	1.50–2.00 each

Light

Equipment	Types and uses	Safety and care	Approximate cost/£
Acetate sheets (coloured)	• These can be cut up to create similar effects to colour paddles.	• Children should use carefully to avoid scratching the surface or bending the acetate.	4–10 per pack
Colour paddles	• Paddle-shaped pieces of acetate for mixing colours or viewing objects.	• Children should use carefully to avoid scratching the surface.	3–6 per set
Concave/convex mirrors	• Double-sided with one side convex and the other concave, producing different reflections. • Can be purchased in different sizes.	• Children should use carefully to avoid scratching the surface.	1.50–7.00 per pack of 10 depending on size

Light

Equipment	Types and uses	Safety and care	Approximate cost/£
Glass mirrors	• Glass mirrors provide better definition than plastic mirrors, which are easily scratched. • Glass mirrors are useful for kaleidoscopes, periscopes and reflecting light from torches.	• Glass mirrors should be backed with sticky plastic or adhesive tape and edges bound with masking tape. This prevents the glass splintering and flying if dropped. • Children should be taught to use glass mirrors with care.	3–9 per pack
Kaleidoscope	• There are different kinds of kaleidoscopes: some have material within the tube to create patterns; others reflect the surrounding environment into the kaleidoscope to create patterns.	• Handle with care; do not drop.	5–15
Lenses	• Lenses with a range of focal lengths needed.	• Children should handle glass lenses with care in order not to scratch the surface or break the lens.	1.50–2.00 each
Light ray box	• Used to illustrate that light travels in straight lines.	• Bulb in box can become hot to touch. • Some require batteries or power pack for use.	10–30
Periscope	• These can be purchased either ready-made or as self-assembly packs. • Large plastic periscopes can also be used under water for pond studies; see *Aqua Scope*.	• Use with care to avoid dislodging mirrors.	8–15
Plastic mirrors	• Plastic mirrors can be bent (to produce concave and convex mirrors) or cut.	• Safe to use, although corners can be sharp. • Plastic is easily scratched and does not produce the same quality of reflection as glass mirrors. • However, they are safer to use than glass mirrors.	3–10 per pack depending on size

Light/Materials and change

Equipment	Types and uses	Safety and care	Approximate cost/£
Prisms	• Prisms can be made from glass or acrylic. • Used to create rainbow effect. • Glass prisms produce a better effect than acrylic.	• Glass prisms can chip. • Best effect is produced by a glass prism on an OHP surface.	4–25 per set
Torches	• It is worth paying extra for good quality torches which produce a strong beam. • Check the torch beam; if it is too weak it can ruin an activity.	• If batteries are left in torches too long they can leak. • If not used on a regular basis remove batteries from torches. • Check that there are batteries for torches to avoid the frustration of not being able to carry out light activities.	2–12 each

Materials and change

Equipment	Types and uses	Safety and care	Approximate cost/£
Droppers (Pasteur pipettes)	• Disposable plastic pipettes. • Usually sold in packs of 100 to 500	• Do not use for cookery.	10–15 per pack
Filter papers	• Used for work on filtering, e.g. soils. • Can be purchased in a range of sizes from 7 cm to 15 cm diameter in packs of 100.	• Store in dry place.	1.50–5.00 per pack
Hob unit (tabletop)	• Portable electric hob suitable for use in the classroom.	• Should only be used by the teacher or under strict adult supervision. • See the ASE book *Be safe!* • Heat-resistant mats should also be used. • Check employer's safety guidelines.	30–50
Metal rods	• Range of rods made from different metals for use in activities on properties of metals, including thermal conduction.		4–5

Materials and change

Equipment	Types and uses	Safety and care	Approximate cost/£
Night lights	• Long-burning in foil tray, usually sold in packs of 10.	• See the ASE book *Be safe!* • Only use under strict adult supervision where children are able to follow safety procedures. • Use a metal baking tray with dry sand in the base and the night light (or candle) in the middle.	1–2
Pestle and mortar	• Used for crushing. • Available in different sizes.	• Made from unglazed porcelain. Care should be taken not to drop and break.	3–9
Retort stand	• Used for holding items in an activity. • Requires the base, rod, bosshead and clamp for a complete retort stand.	• Ensure components are fixed securely. • Made from heavy cast iron which can cause injury if dropped.	6–10
Safety spectacles	• Available for children from 4 years of age upwards. • Safety spectacles can be purchased which are adjustable and have side shields.	• Safety spectacles should be used where there is the danger of pieces flying into the eye. • Only purchase safety spectacles carrying a British (BS2092) or new European (EN166) Standard number. • Try to avoid activities where safety spectacles might be required. • See the ASE book *Be safe!* for further information.	2–5
Specimen (test-) tube rack	• To hold specimen (test-) tubes.		2–4
Tongs	• For holding specimens (e.g. pieces of fabric) and test-tubes over heat source (e.g. night light).	• See the ASE book *Be safe!*	1–2

Measuring equipment: Magnifiers and viewers

Equipment	Types and uses	Safety and care	Approximate cost/£
Aqua Scope	• Periscope which can be used underwater for viewing in ponds and rock pools.	• Safety when children are working near and in water.	13–20
Fresnel lens	• Plastic lens varying from a credit-card size to 30 cm by 30 cm.	• As with all magnifiers children should use with care to avoid scratching the surface.	1–5
Flexi-magnifier	• Lens mounted on a metal stand for stability which allows the arm and therefore the lens to move.	• As with all magnifiers children should use with care to avoid scratching the surface.	20–30
Hand magnifiers	• Plastic hand lenses in a range of sizes.	• Can be scratched. • As with all magnifiers should be stored with care.	2–4
Magnispector	• Transparent container with magnifying lid. • The container has a metric grid printed on the base.	• Do not place stones or other sharp objects in the magnispector since they can damage the container. • Store carefully and ensure that the lens lid is always returned with the container.	10–17
Midispector	• Similar to a magnispector but smaller.	• Do not place stones or other sharp objects in the midispector since they can damage the container. • Store carefully and ensure that the lens lid is always returned with the container.	10
Minispector	• Smaller version of magni- and midispectors without the grid on the base. • Used as a sample box for viewing small animals.	• Do not place stones or other sharp objects in the minispector since they can damage the container. • Store carefully and ensure that the lens lid is always returned with the container.	2–3

Measuring equipment: Magnifiers and viewers

Equipment	Types and uses	Safety and care	Approximate cost/£
Monocular microscope	• 'Stereo' style microscope with one fixed eyepiece. • A useful introductory microscope for children. • Can be used for viewing large objects. • Do not be tempted to buy 'toy' microscopes.	• Never let children direct a microscope mirror towards the Sun; the reflected rays could damage their eyes. • A table lamp provides a safe light source.	45–55
Naturescope or bugscope	• Container with magnifiers which looks like a transparent set of binoculars. • Used to study pond and insect life.	• Contains small parts, such as lids. Ensure that these pieces are not lost.	8–10
Nature viewers	• Small containers with lens lid. • For use studying insects and pond life.	• Count out and count in; these are easily lost. • Ensure that the lens lid is always returned with the container.	1–2
Pocket lenses	• Small magnifier in a case.	• Count out and count in; these are easily lost.	2–4
Pocket hand microscope	• Illuminated hand-held microscope. • Excellent for viewing fabrics, wood, etc., or surfaces of objects.	• Batteries are required. • Children will need to be taught how to use this microscope.	5–10
Stereo microscope	• Both eyes are used to view the object under the microscope. • Easier to use than a monocular microscope. • Designed for viewing objects in 3D.	• Never let children direct a microscope mirror towards the Sun; the reflected rays could damage their eyes. • A table lamp provides a safe light source.	70–90

Measuring equipment: Magnifiers and viewers/Thermometers

Equipment	Types and uses	Safety and care	Approximate cost/£
Table-top magnifier	• Lens which adjusts to any angle on an axis. • Useful for activities which require hands to be free.	• Store carefully. • Do not use to direct Sun's rays.	10–15
Tripod magnifier	• Lenses mounted on a tripod with different magnifications.	• Store carefully.	16–20
Two-way microscope	• This allows children to view from the top and the bottom of the viewer. • Useful for allowing children to view the underside of invertebrates.	• Store with care so that the parts of the microscope do not get lost. • Avoid sharp objects that can scratch the plastic.	4–10

Measuring equipment: Thermometers

Equipment	Types and uses	Safety and care	Approximate cost/£
Digital clinical thermometer	• Suitable for work on 'Ourselves' • Can be used under-arm to take body temperature. • Has an automatic 'switch off' to conserve battery.	• These are small and easily lost. Count out and in.	5–8
Digital environmental thermometer	• Suitable for indoor and outdoor use, e.g. pond work.	• Count out and count in to ensure the thermometer is not left outdoors and lost.	10–15
Digital thermometer	• Digital thermometer with probe suitable for use in liquids.	• Probe must not be forced into material. • Count out and count in.	10–25

Measuring equipment: Thermometers

Equipment	Types and uses	Safety and care	Approximate cost/£
Forehead strip thermometer	• Useful for work on 'Ourselves' with very young children. • Use on forehead to take body temperature.		10–15 per pack
Giant wall thermometer	• Very large wall thermometer. • Suitable for very young children. • Some have interchangeable scales including pictures, words and plain strips so that teacher or children can write or draw pictures.	• As with all wall thermometers, do not place in direct sunlight.	12–18
Large wall thermometer		• As with all wall thermometers, do not place in direct sunlight.	3–5
Liquid crystal temperature indicator sheet	• Flexible plastic sheet which has a liquid crystal layer. The sheet changes colour with changes in temperature. • Can be divided to make smaller LCD thermometers.	• Temperature range of individual sheets may be limited. Check the range of the sheet being purchased.	30
Maximum/ minimum thermometer	• Indicates maximum and minimum temperatures.	• Requires re-setting. • As with all wall thermometers, do not place in direct sunlight.	5–8

Measuring equipment: Thermometers/Length

Equipment	Types and uses	Safety and care	Approximate cost/£
Spirit thermometer (glass stirring thermometer)	• Used for a range of activities, e.g. insulation, dissolving. • These can be purchased in different sizes. • Liquids vary in colour, e.g. red or green. • Children should be taught how to read a thermometer.	• These thermometers do break, although they snap into two rather than shatter into small parts. • Spirit thermometers, although less accurate than mercury, should be used because it is difficult to clear up a mercury spillage. • Children should be taught how to handle thermometers if they break, and not to put broken glass in the class waste paper bin. • See the ASE book *Be safe!*	1–2
Thermostik	• Virtually unbreakable. • Can be used in soil and dipped into ponds.	• Although very sturdy does still need to be handled with care, particularly if being used in soil.	10–15
Window thermometer	• An external thermometer placed on the outside of a window and viewed from inside.	• The reading will be affected if placed in direct sunlight.	3–5

Measuring equipment: Length

Equipment	Types and uses	Safety and care	Approximate cost/£
Callipers	• Callipers allow the measurement of different objects, e.g. rocks, foot size.	• If using to measure children, it is important to develop sensitive attitudes within the class in relation to differences between children.	5–17
Height chart	• Attractive wallcharts against which children can measure their height.	• If using to measure children, it is important to develop sensitive attitudes within the class in relation to differences between children.	8–20
Height measurer	• Free-standing measure which can be used to measure the height of children as well as the height of chairs, desks, etc.	• If using to measure children, it is important to develop sensitive attitudes within the class in relation to differences between children.	25–30

Measuring equipment: Length/Volume

Equipment	Types and uses	Safety and care	Approximate cost/£
Metre sticks	• Both wooden and plastic are available.	• Plastic metre sticks are durable and children can use them in stream or pond work.	3–6
Tape measures	• Range from basic tailor's tape to a retractable tape measure in a plastic case. • Children should appreciate fitness for purpose and recognise which type of tape measure is best for the activity.	• Always roll tapes up after use. Store rolled to avoid damage.	2–4 per pack of 5 tailor's tapes 8–25 cased tapes
Trundle wheel	• Wheel which clicks to signify when a distance of 1 m has been covered.		10–15

Measuring equipment: Volume

Equipment	Types and uses	Safety and care	Approximate cost/£
Measuring beakers	• Plastic beakers in different sizes from 50 cm^3 to 1000 cm^3.	• Store together; they are usually sold as a set. • It is frustrating to find one particular beaker has gone missing. Check out and in.	0.50–2.00 each 5–15 per set
Measuring cylinders	• Plastic measures in different sizes from 10 cm^3 to 1000 cm^3.	• Store together; they are usually sold as a set. • It is frustrating to find one particular cylinder has gone missing. Check out and in.	0.60–5.00 each 20–25 per set
Measuring jugs	• Jugs for pouring in a range of sizes.	• Do not use any science equipment for cookery; that includes measuring jugs.	7–10

Measuring equipment: Volume/Time

Equipment	Types and uses	Safety and care	Approximate cost/£
Measuring spoons	• Set of plastic spoons usually on ring. • Capacity 1, 2, 5, 15, and 25 ml (cm^3)	• Do not use any science equipment for cookery; that includes measuring spoons.	1–5
Syringes	• A range of sizes is required. • Useful for measuring small amounts of water. • Used for giving plants set amounts of water; children find them easier to use than pouring from measuring beakers. • Also useful for teaching forces, either pneumatics or hydraulics.	• Only purchase from educational suppliers. • Accept only syringes in sealed packages.	1–3 each 25–30 for pack of 100

Measuring equipment: Time

Equipment	Types and uses	Safety and care	Approximate cost/£
Digital stopwatches	• Vary in size and range and complexity of functions. • Simpler versions are easier for children to use.	• Children often have difficulty with decimal point and writing the correct time from the watch.	3–10
LCD stopclock	• Digital display with stop/start facilities.	• If left for long periods, take out the battery.	10–15
Mechanical stopclock	• Analogue face with second and minute hands. • Has stop/start facilities. • One of the few clocks which 'ticks' and therefore can be used in sound investigations.	• Do not bang or drop. Use with care.	15–20
Mechanical stopwatch	• Mechanical internal mechanism. • Analogue face.	• Use with care. Do not overwind.	13–16

Measuring equipment: Time/Weight/mass

Equipment	Types and uses	Safety and care	Approximate cost/£
Sand timer	• Suitable for young children. • Different running lengths from 30 seconds to 5 minutes are available.	• Although these are robust, children should be taught that, as with all equipment, timers should be handled with care.	5

Measuring equipment: Weight/mass

Equipment	Types and uses	Safety and care	Approximate cost/£
Bathroom scales	• For use in topics such as 'ourselves' for measuring weight as well as strength of arms.	• If using to weigh children, it is important to develop sensitive attitudes within the class in relation to differences between children.	11–15
Bucket balances	• For younger children. • Some have transparent buckets.	• Store carefully to avoid damage.	15–30
Electronic scales	• These have LCD display. • Useful for small amounts of material.	• Requires batteries. • If left for long period batteries should be taken out. • Do not overload.	15–30
Kitchen scales	• Usually have top pan. • Pointer may need to be reset before use. • Choose scales with clear dial face.	• Do not overload. • Scales used for science should not be used for cookery.	7–20
Metal masses	• Steel or brass masses. • Steel masses have ring which can be attached to pulleys, etc.	• Heavier masses can cause injury if dropped on fingers or toes. • See the ASE book *Be safe!* for details on how to use masses safely.	3–10
Newton meters	• Measure pull force in newtons and g. • Colour-coded for convenience. • Most are in a clear plastic casing so that spring mechanism is clearly visible and children can see how a Newton meter works.	• Children should be taught that there is a range within Newton meters and that the correct one for the job should be chosen. • Do not over-stretch spring.	4–6 each

Measuring equipment: Weight/mass

Equipment	Types and uses	Safety and care	Approximate cost/£
Plastic masses	• Usually sold in sets.	• Keep together to avoid losing masses.	10–40 per set
Slotted masses	• Slotted masses and hangers. • Masses slot on to the hanger and should not drop off accidentally. • Useful to test the strength of, for example, thread or the stretch of an elastic band or fabric. • Hangers and masses can be bought separately or as a set.	• Do not let hanger drop on to a hard surface; it can break. • Masses can be dangerous if dropped on fingers or toes. • See the ASE book *Be safe!*	12–25 per set 1.50–20.00 for individual pieces

Please note

For children with sight impairment there are suppliers who offer measuring equipment with Braille.

Equipment audit list

Name	Number	Comments
The environment		
anemometer		
aquarium		
barometer		
binoculars		
compass (magnetic)		
flower press		
gardening tools		
spade		
Dutch hoe		
fork		
rake		
hand trowel		
hand fork		
secateurs		
watering can		
Garden spray		
wheelbarrow		
globe		
hygrometer		
light meter/sensor		
litmus paper		
nets		
pH meter/probes		
pH papers		
plant propagators		
pooters		

Name	Number	Comments
quadrats		
rocks and fossils		
rain gauge		
soil meters		
moisture		
pH		
temperature		
soil sieves		
weather board		
Ourselves		
anatomical models		
ear		
eye		
heart		
human anatomy		
human torso		
inside-out tunics		
skeleton		
tooth		
dental charts		
heartbeat counter		
stethoscope		
X-rays		
Sound		
musical instruments		
plastic tubing		
plastic funnels		

Equipment audit list

Name	Number	Comments	Name	Number	Comments
pipes (metal)			**Magnets**		
slinky			Alnico magnets		
sound-level meter			bar/horseshoe magnets		
tuning forks			ceramic magnets		
Electricity			floating (ring or polo) magnets		
batteries (dry cell)			horseshoe magnets (infant)		
battery holders			iron filings		
bells			iron filings bubble		
bulbs (lamps)			magnetic marbles		
bulb holders (lamp holders)			magnetic rubber strips and sheets		
buzzers			magnetic tape		
crocodile clips			magnetic wands		
LEDs (light emitting diodes)			metal discs		
motors			soft iron rods		
screwdrivers			**Light**		
switches *bell push press/push reed slide on/off toggle throw tilt*			acetate sheets (coloured)		
			colour paddles		
			concave/convex mirrors		
			glass mirrors		
			kaleidoscope		
			lenses		
wire			light ray box		
wire strippers					

Equipment audit list

Name	Number	Comments
periscope		
plastic mirrors		
prisms		
torches		

Materials and change

Name	Number	Comments
droppers (Pasteur pipettes)		
filter papers		
hob unit (tabletop)		
metal rods		
night lights		
pestle and mortar		
retort stand		
safety spectacles		
specimen (test-) tube rack		
tongs		

Measuring equipment: Magnifiers and viewers

Name	Number	Comments
Aqua Scope		
Fresnel lens		
flexi-magnifier		
hand magnifiers		
magnispector		
midispector		
minispector		

Name	Number	Comments
monocular microscope		
naturescope or bugscope		
nature viewers		
pocket lenses		
pocket hand microscope		
stereo microscope		
table-top magnifier		
tripod magnifier		
Two-way microscope		

Measuring equipment: Thermometers

Name	Number	Comments
digital clinical thermometer		
digital environmental thermometer		
digital thermometer		
forehead strip thermometer		
giant wall thermometer		
large wall thermometer		
liquid crystal temperature indicator sheet		
maximum/ minimum thermometer		
spirit thermometer (glass stirring thermometer)		
Thermostik		
window thermometer		

Equipment audit list

Name	Number	Comments
Measuring equipment: Length		
callipers		
height chart		
height measurer		
metre sticks		
tape measures		
trundle wheel		
Measuring equipment: Volume		
measuring beakers		
measuring cylinders		
measuring jugs		
measuring spoons		
syringes		
Measuring equipment: Time		
digital stopwatches		
LCD stopclock		
Mechanical stopclock		
mechanical stopwatch		
sand timer		
Measuring equipment: Weight/Mass		
bathroom scales		
bucket balances		
electronic scales		
kitchen scales		

Name	Number	Comments
metal masses		
Newton meters		
plastic masses		
slotted masses		
Other equipment		

Small equipment and consumables

Equipment	Number	Comments
aluminium foil		
balloon pump		
balloons		
balls		
beads		
bricks		
bubble makers		
candles		
card offcuts		
card tubes		
clothes pegs		
coat hangers – metal		
combs		
compost		
corks		
cups – clear plastic		
cotton reels		
disposable gloves		

Equipment	Number	Comments
dowel		
elastic bands		
fabric collection		
fishing line		
glass paper		
guitar strings		
guttering		
hand fans (battery operated)		
heat-proof mats		
magnetic toys		
marbles		
metal collection		
nails		
paper fasteners		
paper clips		
plant pots		
plastic bags		
plastic collection		

Small equipment and consumables

Equipment	Number	Comments
plastic containers		
Plasticine		
rubber bungs		
seed trays		
seeds/pulses for shakers		
shirring elastic		
sponges		
spoons – plastic, metal and wooden		
springs		
straws – paper and plastic		
string		
thread		
washing-up bottles		
washers		
wind-up toys		
wood collection		
wood offcuts		
wool collection		

Equipment	Number	Comments
ADDITIONAL ITEMS		

ASE • PRIMARY SCIENCE EQUIPMENT 49

Photocopiable equipment labels

- The following pages can be photocopied and enlarged or reduced.
- Use the labels on boxes, shelves, trays, etc., to help locate equipment.

- The small squares on each label can be used to record numbers of individual items.

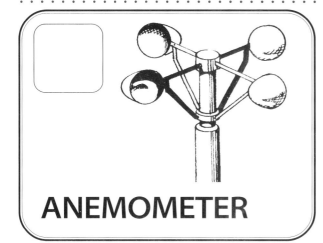

ANEMOMETER

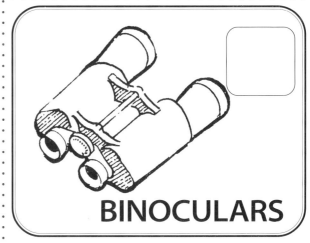

BINOCULARS

AQUARIUM

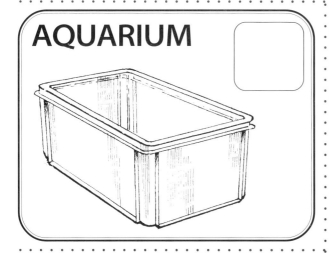

COMPASS (MAGNETIC)

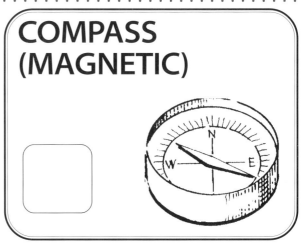

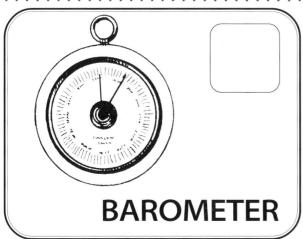

BAROMETER

FLOWER PRESS

GARDENING TOOLS

LITMUS PAPER

GLOBE

NETS
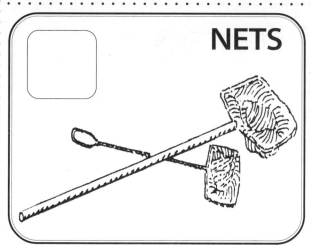

HYGROMETER
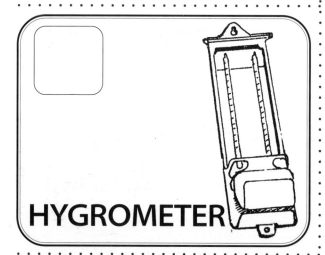

pH METER/PROBES
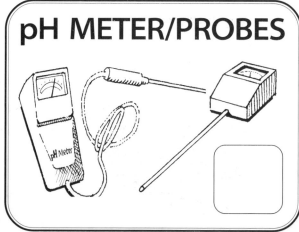

LIGHT METER/SENSOR
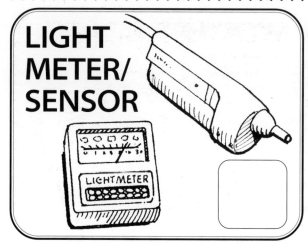

pH PAPERS

PLANT PROPAGATORS
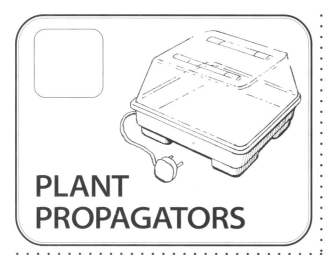

RAIN GAUGE
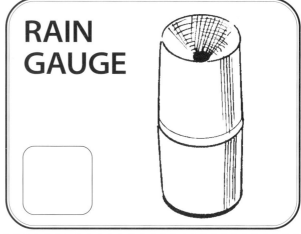

POOTERS

SOIL METERS
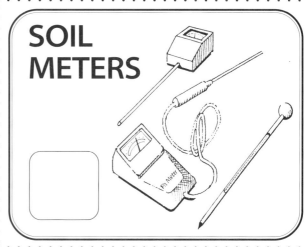

QUADRATS
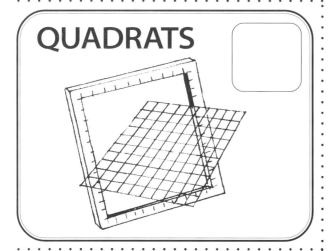

SOIL SIEVES

ROCKS AND FOSSILS

WEATHER BOARD
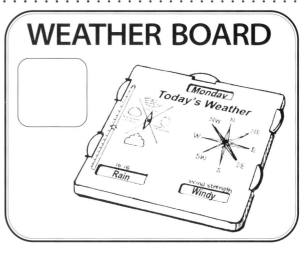

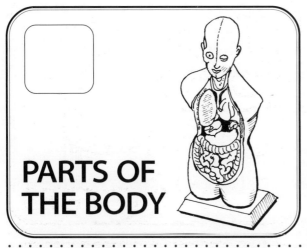
PARTS OF THE BODY

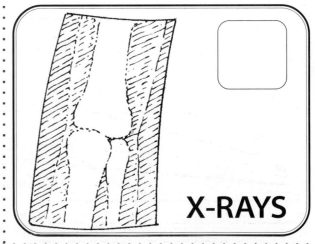
X-RAYS

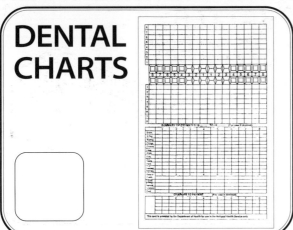
DENTAL CHARTS

MUSICAL INSTRUMENTS

HEARTBEAT COUNTER

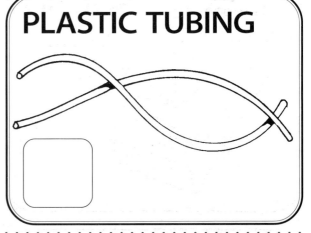
PLASTIC TUBING

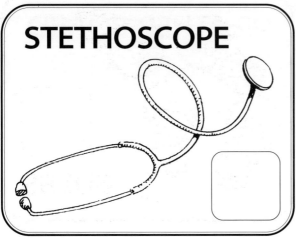
STETHOSCOPE

PLASTIC FUNNELS

ASE • Primary Science Equipment 53

PIPES
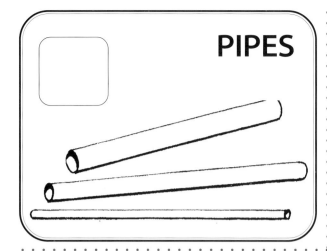

BATTERIES
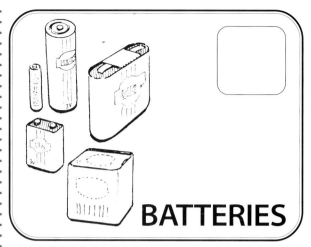

SLINKY

BATTERY HOLDERS
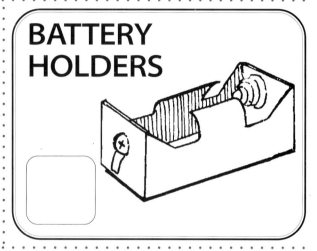

SOUND LEVEL METER

BELLS

TUNING FORKS
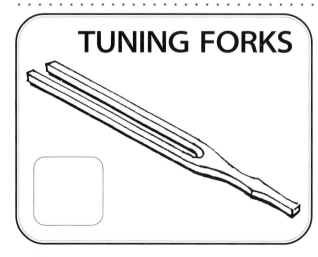

BULBS

BULB HOLDERS

MOTORS

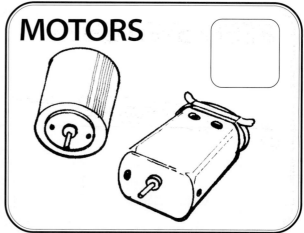

BUZZERS

SCREWDRIVERS

CROCODILE CLIPS

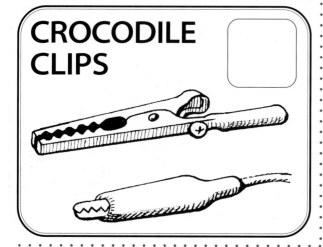

BELL PUSH SWITCHES

LEDs

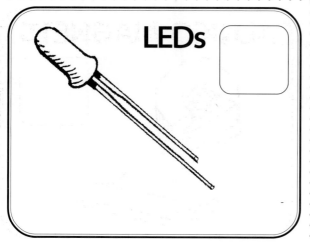

PRESS/PUSH SWITCHES

REED SWITCHES
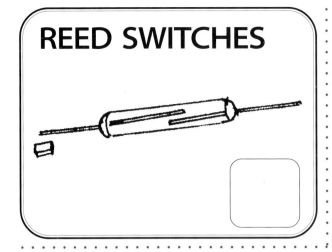

TILT SWITCHES

SLIDE ON/OFF SWITCHES

WIRE
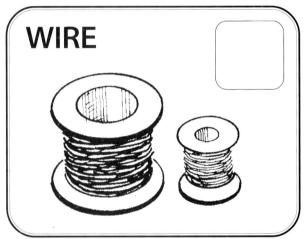

TOGGLE SWITCHES

WIRE STRIPPERS

THROW SWITCHES
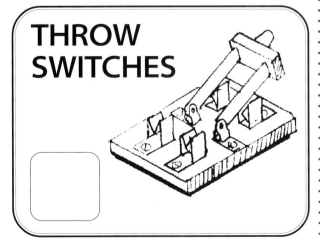

ALNICO MAGNETS
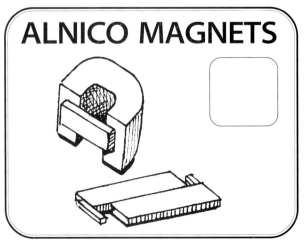

BAR/ HORSESHOE MAGNETS
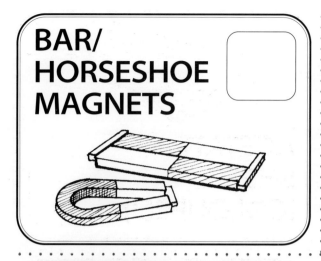

IRON FILINGS
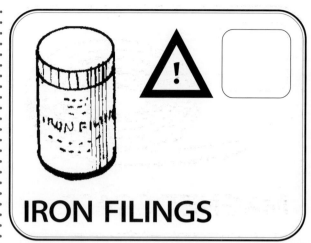

CERAMIC MAGNETS
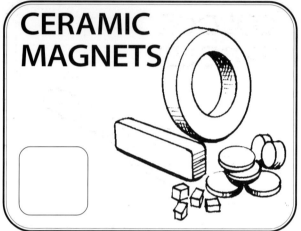

IRON FILINGS BUBBLE

FLOATING (RING OR POLO) MAGNETS
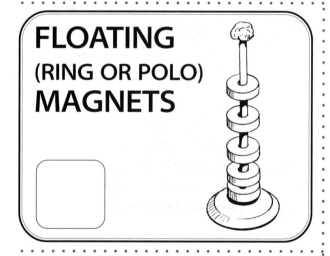

MAGNETIC MARBLES

HORSESHOE MAGNETS (INFANT)

MAGNETIC RUBBER STRIP/SHEET
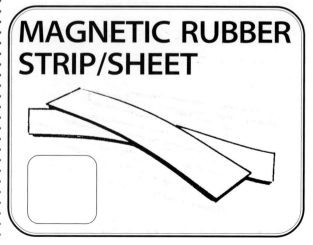

ASE • Primary Science Equipment 57

MAGNETIC TAPE
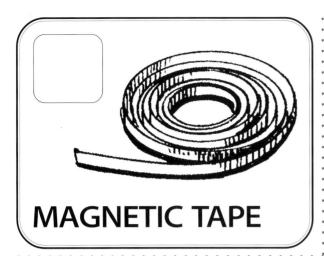

ACETATE SHEETS (COLOURED)
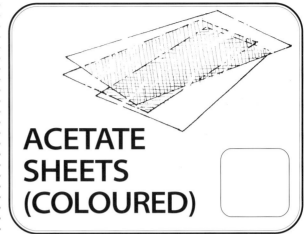

MAGNETIC WANDS
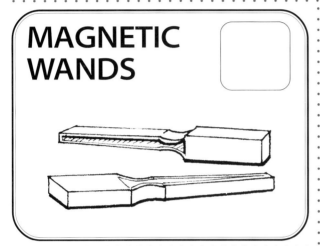

COLOUR PADDLES

METAL DISCS
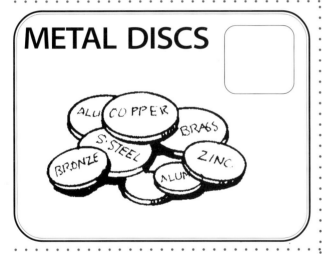

CONCAVE/CONVEX MIRRORS
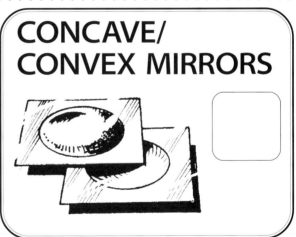

SOFT IRON RODS
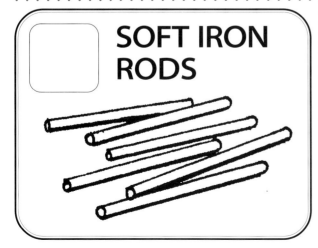

GLASS MIRRORS

KALEIDOSCOPE

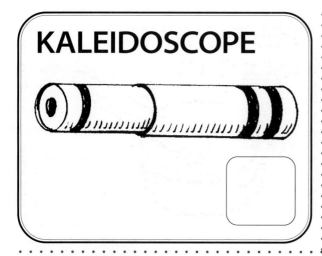

PLASTIC MIRRORS

LENSES

PRISMS

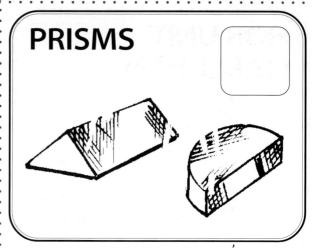

LIGHT RAY BOX

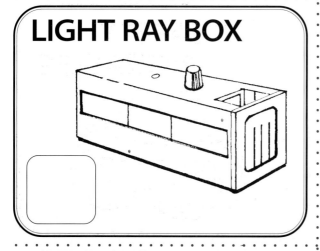

TORCHES

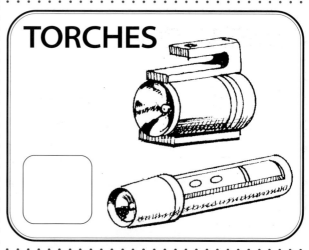

PERISCOPE

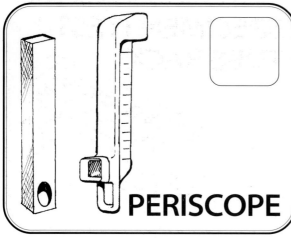

DROPPERS (PASTEUR PIPETTES)

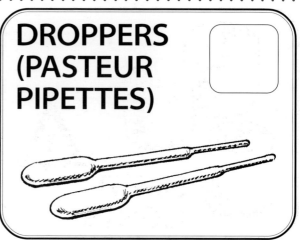

FILTER PAPERS

PESTLE AND MORTAR
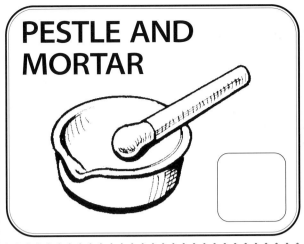

HOB UNIT (TABLETOP)

RETORT STAND

METAL RODS
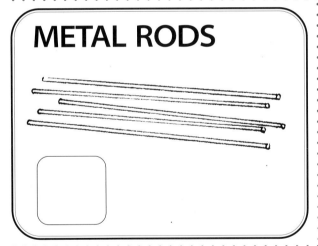

SAFETY SPECTACLES

NIGHT LIGHTS

SPECIMEN (TEST-) TUBE RACK
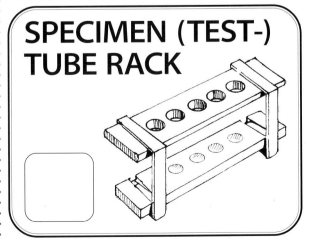

TONGS 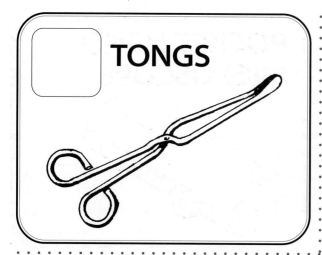	**HAND MAGNIFIERS** 
AQUA SCOPE 	**MAGNISPECTOR** 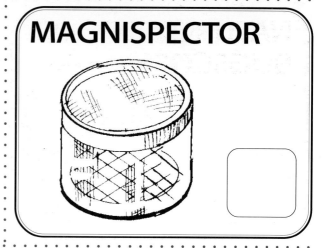
FRESNEL LENS 	**MIDISPECTOR** 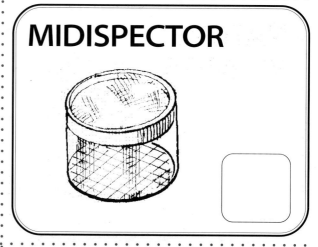
FLEXI-MAGNIFIER 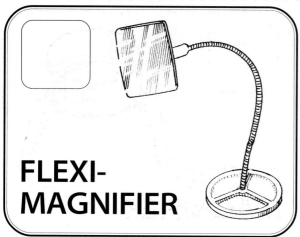	**MINISPECTOR** 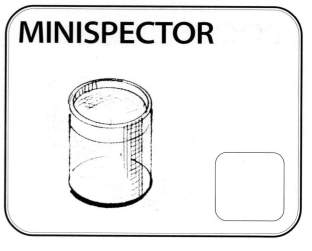

ASE • PRIMARY SCIENCE EQUIPMENT 61

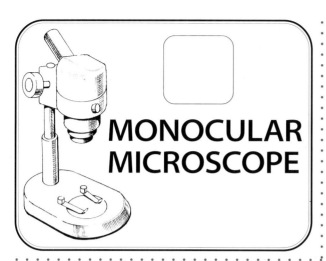

MONOCULAR MICROSCOPE

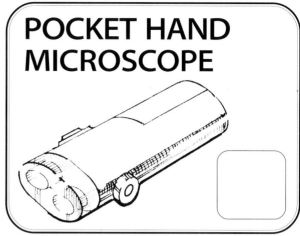

POCKET HAND MICROSCOPE

NATURESCOPE OR BUGSCOPE

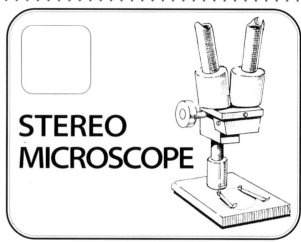

STEREO MICROSCOPE

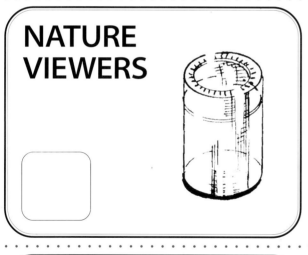

NATURE VIEWERS

TABLE-TOP MAGNIFIER

POCKET LENSES

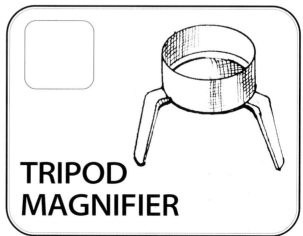

TRIPOD MAGNIFIER

TWO-WAY MICROSCOPE

FOREHEAD STRIP THERMOMETER

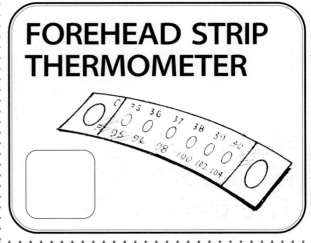

DIGITAL CLINICAL THERMOMETER

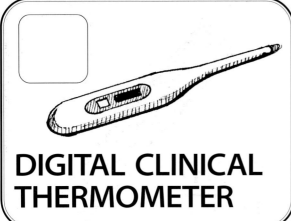

GIANT WALL THERMOMETER

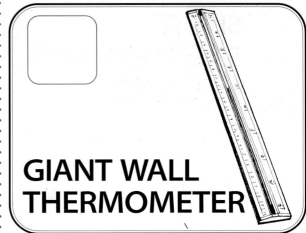

DIGITAL ENVIRONMENTAL THERMOMETER

LARGE WALL THERMOMETER

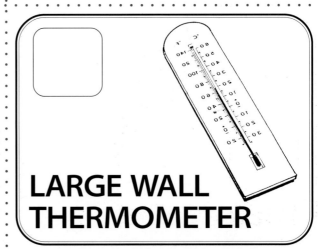

DIGITAL THERMOMETER

LIQUID CRYSTAL TEMPERATURE INDICATOR SHEET

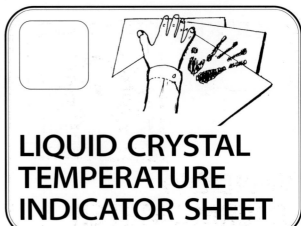

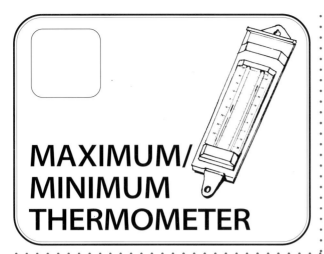
MAXIMUM/ MINIMUM THERMOMETER

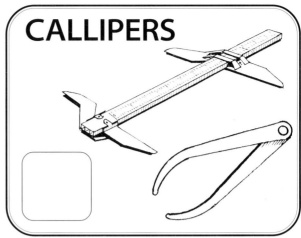
CALLIPERS

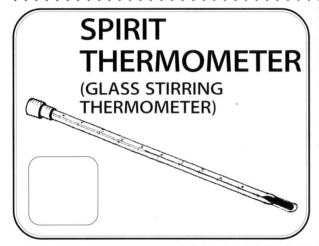
SPIRIT THERMOMETER
(GLASS STIRRING THERMOMETER)

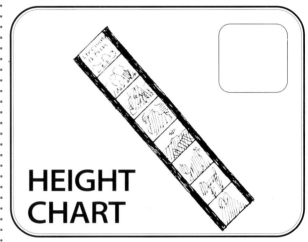
HEIGHT CHART

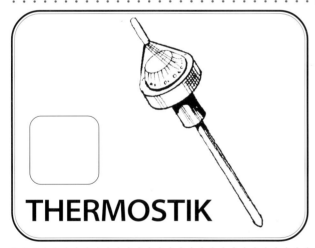
THERMOSTIK

HEIGHT MEASURER

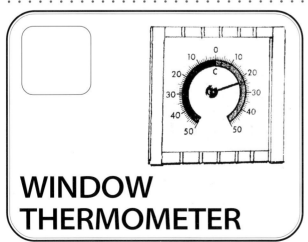
WINDOW THERMOMETER

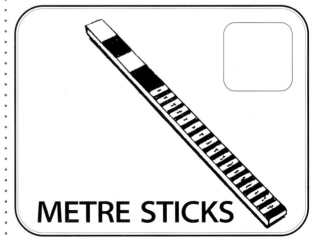
METRE STICKS

TAPE MEASURES
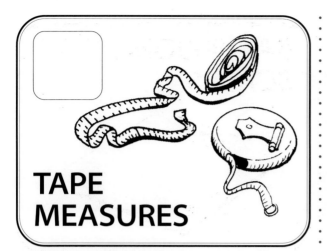

MEASURING JUGS
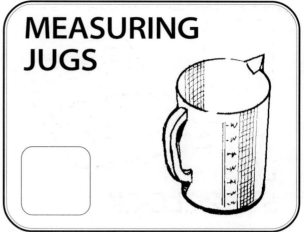

TRUNDLE WHEEL
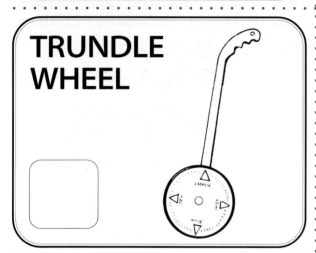

MEASURING SPOONS

MEASURING BEAKERS
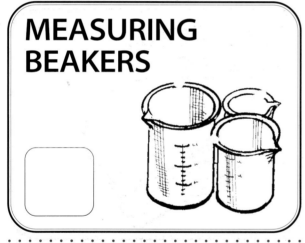

SYRINGES
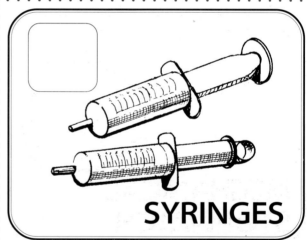

MEASURING CYLINDERS
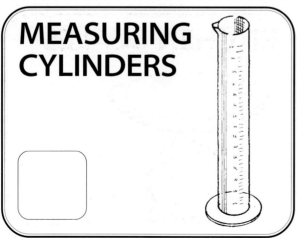

DIGITAL STOPWATCHES
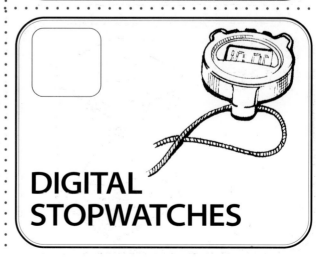

LCD STOPCLOCK
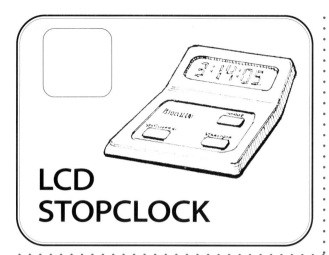

BATHROOM SCALES

MECHANICAL STOPCLOCK

BUCKET BALANCES

MECHANICAL STOPWATCH
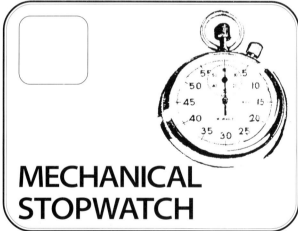

ELECTRONIC SCALES
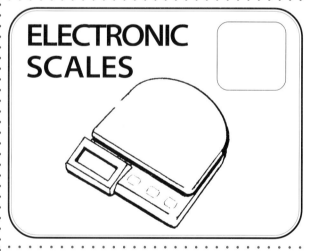

SAND TIMER
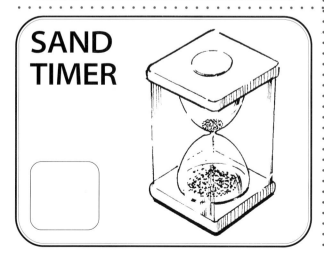

KITCHEN SCALES

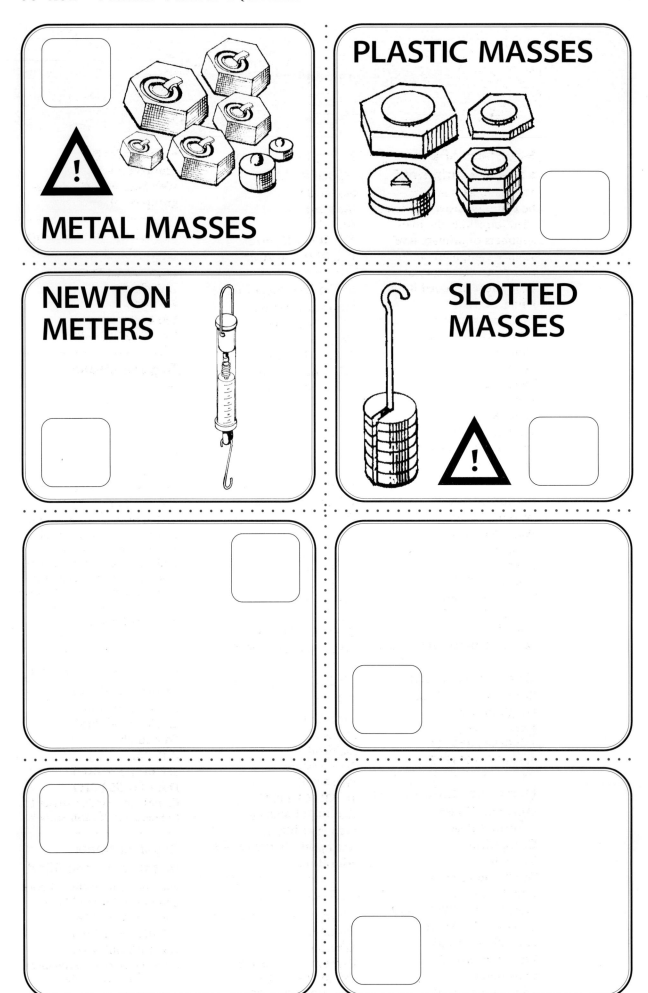

Useful addresses

Suppliers of primary science equipment

This list makes no claims to be comprehensive. We have tried to include the main suppliers of primary level science equipment and to cover the main types of equipment included in this guide.

Commotion
Unit 11
Tannery Road
Tonbridge
Kent
TN9 1RF
Tel: 01732 773399
Fax: 01732 773390

Data Harvest Educational
Woburn Lodge
Waterloo Road
Linslade
Leighton Buzzard
Beds LU7 7NR
Tel. 1525 373666
Fax: 1525 851638
E-mail: sales@dharvest.demon.co.uk

Griffin and George Ltd
Bishop Meadow Road
Loughborough
Leics L11 5RG
Tel: 01509 233344
Fax: 01509 231893

Heron Educational Ltd
Carrwood House
Carrwood Road
Chesterfield
S41 9QB
Tel: 01246 453354
Fax: 01246 260876
Freephone: 0800 373249
E-mail: sales@heron-educational.co.uk
Web site: http://www.heron-educational.co.uk

Hogg Laboratory Supplies Ltd
Sloane Street
Birmingham
B1 3BW
Tel: 0121 233 1972
Fax: 0121 236 7034

Irwin-Desman Ltd
Eurocrown House
23 Grafton Road
Croydon
Surrey CR9 3AZ
Tel: 0181 686 6441
Fax: 0181 681 8429

Locktronics Ltd
Unit 11
Chancellors Pound
Redhill
Avon BS18 7TZ
Tel: 01934 863560
Fax: 01934 863561

NES Arnold Ltd
Ludlow Hill Road
West Bridgford
Nottingham
NG2 6HD
Tel: 0115 971 7700
Free fax: 0500 410420

Philip Harris Education
Lynn Lane
Shenstone
Lichfield
Staffordshire
WS14 0EE
Tel: 01543 480077
Fax: 01543 480068
Web site: http://www.philipharris.co.uk/education

TTS – Technology Teaching Systems Ltd
Monk Road
Alfreton
Derbyshire DE55 7RL
Tel: 01773 830255
Fax: 01773 830325
E-mail: sales@tts-group.co.uk
Web site: http://www.tts-group.co.uk

Yorkshire Purchasing Organisation
41 Industrial Park
Wakefield
WF2 0XE
Tel: 01924 824477
Fax: 01924 834805

Organisations

CLEAPSS School Science Service
Brunel University
Uxbridge
UB8 3PH
Tel: 01895 251496
Fax: 01895 814372
An information, training and advisory service covering all aspects of practical work and equipment for school science. Works on a membership basis (most schools are members via their LEA's subscriptions). Does not cover Scotland.

Scottish Schools Equipment Research Centre (SSERC)
St Mary's Building
23 Holyrood Road
Edinburgh
EH8 8AE
Tel: 0131 558 8180
Fax: 0131 558 8191
Carries out a similar service to CLEAPSS for Scottish schools.

Royal National Institute for the Blind
Education Information Service
224 Great Portland Street
London W1N 6AA
Tel: 0171 388 1266
Fax: 0171 383 4921
Gives guidance on equipment for children with sight impairment.